# NUCLEAR POWER

Nigel Hawkes

# World Issues

Cities in Crisis
Endangered Wildlife
Exploitation of Space
Food or Famine?
Human Rights
International Terrorism
Nuclear Power
Nuclear Weapons
Population Growth
Refugees
Sport and Politics
The Arms Trade
The Energy Crisis
The Environment
The International Debt Crisis
The International Drug Trade
Threatened Cultures
World Health

**Cover picture:** A Greenpeace boat, the *Moby Dick*, sails past Dounreay nuclear power station in Scotland.
**Frontispiece:** Radiation levels are checked by an engineer wearing a protective mask and overalls.

**Editor:** Marcella Streets
**Series Designer:** David Armitage

First published in 1989 by
Wayland (Publishers) Ltd,
61 Western Road, Hove
East Sussex, BN3 1JD, England

© Copyright 1989 Wayland (Publishers) Ltd

**British Library Cataloguing in Publication Data**

Hawkes, Nigel
 Nuclear power.
 1. Nuclear power – For children
 I. Title
 621.48

 ISBN 1-85210-609-3

Phototypeset by Kalligraphics Ltd, Horley, Surrey
Printed and bound in Italy by Sagdos S.p.A., Milan

# Contents

1. Introduction 6
2. Unimagined power 8
3. Taming the dragon 15
4. Nuclear fuel and nuclear ashes 21
5. What can go wrong? 27
6. Is a nuclear future inevitable? 37
Glossary 44
Books to read 45
Further information 46
Index 47

# 1 Introduction

**Chernobyl**

Very late on the night of 25 April 1986 Natasha Timofeyeva, a 16-year-old schoolgirl, was returning home with her family from a visit to friends. They lived in the tiny village of Chamkov, close to the Pripyat river and the border with the Ukraine. It was quite dark, and as Natasha looked out of the window of the car, she saw what she later described as a bright flash. It was immediately above one of the four chimneys of the local power station.

Natasha was one of the few eyewitnesses of the world's worst nuclear accident. It completely destroyed a reactor at Chernobyl power station, held the world in suspense for two weeks as clouds of dangerous radioactivity spread across Europe, and cast a different kind of cloud over the whole future of nuclear power. It also changed Natasha's life for ever.

On the following day militia arrived in the village, which was less than four miles from the power station. Villagers were ordered on to buses in whatever they stood up in. There was no time for the Timofeyevs to change, pack or to collect more than a few valuables. They were not allowed to leave in their own car, for fear of the roads being jammed with panic-stricken people.

Natasha and her family were forced to join 135,000 other people evacuated from the region around the stricken power station. It is unlikely that she will ever be able to return, for in spite of enormous efforts to clean up the contamination caused by the disaster, the areas close to the plant are still uninhabitable. By the time they are made safe – if they ever are – most of the evacuees will have settled in new homes and may not want to go back.

*Families from villages and towns around Chernobyl were evacuated to reduce exposure to radiation. Many will never be able to return to their homes.*

## The debate

The Chernobyl disaster was the most shocking event in the history of nuclear power, but the subsequent controversy was nothing new. Once heralded as the cleanest, cheapest source of electricity, nuclear power has become the centre of a passionate debate during the past 15 years. Its opponents claim that it is neither cheap, nor clean; that it threatens the environment and human health; and that it contributes to a dangerous spread of nuclear weapons throughout the world. Its supporters deny these claims. They say that nuclear power has set new standards in safety, both for the public and those who work in the plants, and that without it there will not be sufficient energy supplies to last into the next century. The spread of nuclear weapons is a serious problem, they admit, but it has little to do with the peaceful uses of nuclear power.

There seems to be no meeting point between these two viewpoints. Those who are not specialists find themselves confused, alternately impressed by one set of arguments and then the other.

Exactly the same sense of bewilderment can be seen all over the world. Two nations with similar histories and cultural traditions, France and Italy, reflect the great divide. France, with few domestic coal resources and no oil, has worked frantically to develop nuclear power. Just over the border in Italy a different philosophy prevails. The two operational plants were closed down following a referendum and no more are planned.

If two democratic countries can disagree as profoundly as this, it is hardly surprising that ordinary people remain confused. Yet if the right decisions are to be taken in the future, we cannot leave the subject to the experts. The stakes are far too high, as the experience of Natasha Timofeyeva and her family showed.

Nuclear power is complex, but it is not incomprehensible. The basic concepts can be understood without a degree in nuclear physics, and an understanding must be achieved if the voters of the 1990s are to make informed judgements about whether or not nuclear power should continue. This book will explore and explain the issues without taking sides. But first we will have to look back to the scientists of the nineteenth century, who laid the groundwork for the nuclear revolution.

*When leaders of the world met in Toronto, Canada, for an economic summit, in June 1988 they were greeted by anti-nuclear protesters.*

# 2 Unimagined power

## The discovery of the atom

At the end of the nineteenth century, the world of physics was pretty pleased with itself. Many people believed that everything was known that needed to be known, and that from now on it would be merely a matter of perfecting an already sound understanding of the world. They were quite wrong. A series of discoveries in the last twenty years of the century totally changed our conception of the world we live in.

One of the myths that was shattered in that exciting period was our idea of what things are made of. Commonsense suggested that if you took, say, a piece of iron, and divided it in half, and then in half again, then again, and again, you would eventually end up with a tiny fragment of iron too small to see and too small to be divided, however sharp your chisel. This indivisible, solid particle of matter would be an atom, the basic building block upon which great engineering masterpieces like the Forth Bridge or Crystal Palace ultimately depended. A Victorian engineer would have laughed aloud if you had said that his great works were made of atoms that were in reality mostly empty space, yet this actually proved to be the case.

## Radioactivity

The first cracks in the certainty regarding the construction of the world came with the discovery of mysterious rays that were given off by certain atoms. In 1895 Wilhelm Roentgen, a professor at the University of Wurtsburg in Bavaria, had discovered some rays so peculiar that he called them X-rays, X being the term that mathematicians usually give to the unknown. The discovery fascinated everybody, particularly when Roentgen showed how X-rays could be used to display the bones of the human body, passing through the flesh as if it were not there. Stimulated by this discovery, a French scientist, Henri Becquerel, decided to see if some materials he was studying that had the ability to emit light might also emit X-rays. He found to his delight that a salt containing atoms of the metal uranium, potassium uranyl sulphate, did give off rays that would penetrate through paper wrappings and

**Right** *The first X-rays were taken with primitive equipment, by doctors unaware of the dangers of excessive doses of radiation.*

**Below** *Marie Curie discovered radium and invented the word radioactivity to describe its curious behaviour.*

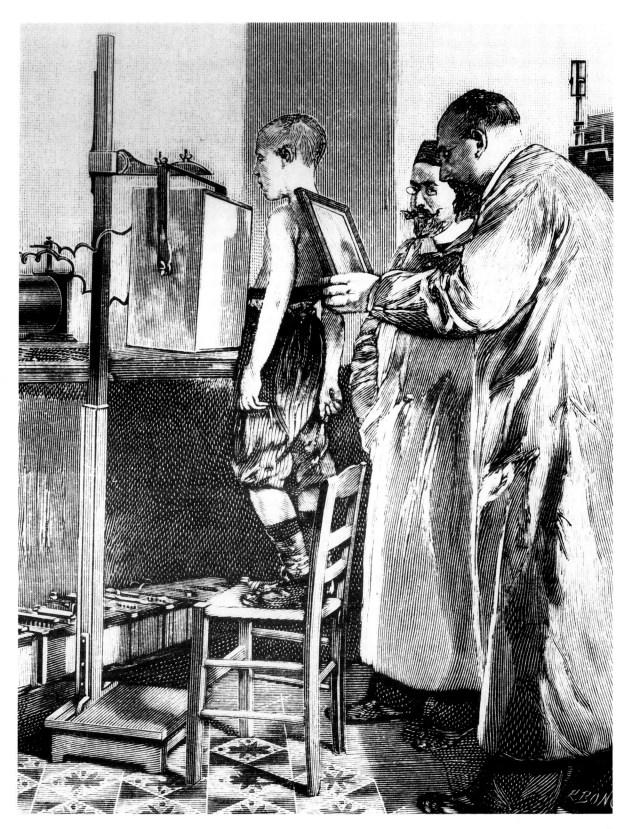

darken photographic plates. They were given off all the time, in every direction, in a continous stream and, like X-rays, they could penetrate some solid objects.

His studies, however, quickly showed that the rays were not X-rays. For a while he called them Becquerel rays until Marie Curie, a young Polish-born scientist, came up with a name that stuck – radioactivity. Becquerel established that the rays could be deflected by a magnetic field – suggesting that the radiation consisted of tiny charged particles – and that they came from uranium atoms. By 1900 he had decided that the rays were in fact tiny speeding particles, identical with those found earlier in cathode ray tubes. They carried a negative electric charge and were called electrons. If electrons were emerging from the atoms of uranium, as they appeared to be, that meant that atoms could certainly not be the solid, billiard ball type of objects scientists supposed them to be. They evidently contained electrons, which could detach themselves and escape.

In the early years of the twentieth century, the number of 'subatomic' particles discovered increased. First the proton was discovered, a positively-charged particle, much bigger than the electron. Then, in 1930, came the first evidence of the neutron, a particle with the same mass as the proton, but no electric charge. The old assumptions of simplicity were shattered. The atom was clearly a complex structure: its kernel, or nucleus, consisted of protons and neutrons, while electrons orbited around it, rather like planets round the sun. It is now known (though it was not then) that the nucleus occupies a mere 1/100,000th of the whole atom. The bulk of the atom is in fact empty space: a difficult concept to grasp, given the undeniable solidity of the world and most of the objects in it.

> I was brought up to look at the atom as a nice hard fellow, red or grey in colour, according to taste.
> *Lord Rutherford, British physicist*

*A neutron hits a uranium atom, triggering fission.*

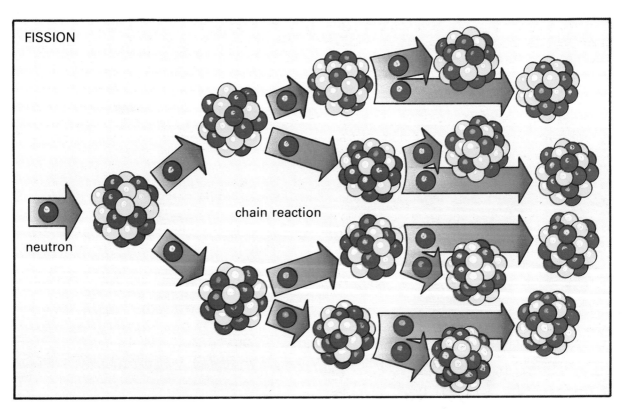

## The splitting of the atom

Meanwhile work had continued with uranium. In 1937 Otto Hahn and Lise Meitner, working in Berlin, made an extraordinary discovery. They found that the nucleus of the uranium atom would itself split if it were bombarded by a stream of neutrons. A single uranium nucleus would divide into two nuclei of lighter elements – the first evidence of the alchemist's dream of 'transmuting' one element into another. What made this discovery important was the minimal effort required to make uranium split; indeed, the process produced a large surplus of energy. And in the course of dividing, the uranium nucleus generated neutrons, which in turn could be used to make other uranium nuclei split. In uranium the whole process, called nuclear fission, was self-sustaining once it had been set in motion, and it generated huge, unprecedented amounts of energy.

In the atmosphere of the late 1930s, this was an alarming discovery. Adolf Hitler and his National Socialist Party were in power in Germany, the country where most of the fundamental discoveries had been made. It was already becoming clear that a war between Nazi Germany and the democracies of France and Britain was inevitable sooner or later, and the two scientists unlocking the door to a process which might be turned into a bomb of unimaginable power were working in Germany.

The plan for making a weapon was as follows. If sufficient uranium could be assembled and the fission reaction started, the process would repeat itself over and over again. One neutron would create one fission, which would produce two neutrons and create two fissions, which would produce four neutrons and four fissions, – and so on. In a fraction of a second the whole mass of uranium would fission, producing a huge blast, intense heat and a withering fusillade of neutrons which would destroy any living thing in range.

All that was needed to make this happen was a sufficiently large lump of uranium. Small pieces would not explode because the neutrons generated would escape into the air before they could cause further fission. Only if a certain mass – known as the critical mass – could be assembled, would a real fission explosion occur. That mass was not too difficult to calculate and it turned out to be around 15 kilograms, about the size of an orange. However, there was more to making a bomb than simply collecting uranium.

Uranium comes in several varieties. Early in the twentieth century it had been shown that pure uranium contained a substance much more radioactive than uranium itself. At first it was called uranium-X, but it was soon demonstrated that the only way in which it differed from the rest of the uranium in which it was found was the number of neutrons in its nucleus. Ordinary uranium (U-238), which makes up more than 99 per cent of uranium found naturally, contains 238 subatomic particles in its nucleus: 92 protons and 146 neutrons. Highly radioactive uranium, only 0.7 per cent of the natural uranium, contains only 235 particles: the same 92 protons, but only 143 neutrons. It is this second form, U–235, which is fissile and which can be turned into a bomb. The different forms of elements caused by variations in the number of particles in the nucleus are known as isotopes.

To make a bomb, then, involved extracting the fissile isotope, U-235, from the rest of the uranium – a very difficult task, because there are no differences in chemical behaviour between the two isotopes which can be used to separate them. There was, though, a second possibility. It was observed that when U-238, the non-fissile isotope, was bombarded with

---

Physics stands at the threshold of discoveries of boundless significance. We are confronted not only with the fact that mankind will acquire a new source of energy surpassing by a million times everything that has been hitherto known ... the central fact is that human might is entering a new era. Man will be able to acquire any quantity of energy he pleases and apply it to any end he chooses.

*Editorial in* Izvestia, *31 December 1940.*

neutrons it was converted into something else entirely, an element which had never been seen before and which did not seem to occur naturally on earth. This was plutonium-239 (known for short as Pu-239), an isotope which was fissile. So there appeared to be two separate routes to the bomb: the first involved isolating as much U-235 as possible by whatever methods could be devised; the second meant bombarding U-238 with neutrons to produce Pu-239, then extracting the product chemically to use in the manufacture of bombs.

## The first reactor – and the first atom bomb

By 1939 a number of scientists had become aware of the possibilities and were terrified that Hitler would be the first to use this bomb. Every ounce of U-235 that fissioned would produce as much power as 610 tonnes of TNT explosive, an awful prospect for the Allies. A group of physicists persuaded the scientist Albert Einstein to write a letter to President Franklin Roosevelt of the USA, warning of the dangers. Einstein was not a nuclear physicist, but his name carried more weight than any other scientist. He wrote the letter on 2 August 1939 but it was not actually handed to Roosevelt until 11 October, by which time war had broken out in Europe. Two years later, on 6 December 1941, Roosevelt authorized a rushed project to see if a bomb could be built. To avoid suspicion, it was given the boring name of the Manhattan Engineering District Project, or the Manhattan Project for short. Its consequences have changed the world we live in.

The Manhattan Project was a collaboration between scientists from the USA, Britain, France and Canada, as well as exiles from the Axis powers, like the Italian Enrico Fermi. It was Fermi who was given the task of producing Pu-239 by bombarding U-238 isotopes with neutrons. He did so by building the first nuclear pile in an old squash court under the football stadium at the University of Chicago, USA.

The idea was this. It had been found that the neutrons released by fission were more likely to cause further fission if they were slowed down first. The best way of achieving this was by letting them pass through a moderator which would temper their speed.

The Manhattan Project chose graphite, the hard black form of carbon that is used as the lead in ordinary pencils. Reasoning that even ordinary uranium would fission (despite its low proportion of U-235) if only there were enough of it, Fermi built a huge pile of alternate layers of uranium and graphite. There were 57 layers in all, and the pile was 9.14 m wide, 9.75 m long and 6.55 m high. It used six tons of uranium, almost all the known supplies available to them at the time.

*Enrico Fermi, the Italian scientist who led the team that built the first nuclear reactor in Chicago, demonstrates at the blackboard the principles of the nuclear chain reaction.*

Through the middle of the pile were placed rods of cadmium, a material that soaks up neutrons and could therefore be used to control the reaction. The key question was whether a chain reaction could be started, and then controlled and maintained at a steady rate so that the neutrons produced by U-235 fission could convert the inactive U-238 into Pu-239. On 2 December 1942 the cadmium rods were slowly withdrawn and the reaction speeded up. At 3.45 pm it became self-sustaining. The first nuclear reactor was working.

A telegram, deliberately obscure in meaning, was immediately despatched to Washington: 'The Italian navigator has entered the new world,' it read. The Office of Scientific Research and Development replied: 'How are the natives?' Back came the reply: 'Very friendly'.

While Fermi and his team were producing plutonium, a second team was attempting to isolate U-235 at laboratories at Los Alamos, in New Mexico. Eventually both teams did

*At 3.45 pm on 2 December 1942, scientists observed the first controlled nuclear chain reaction in a squash court at the University of Chicago.*

succeed and several bombs were produced. The first was tested on 16 July 1945 at Alamagordo, New Mexico. It produced a stupendous explosion, equal to 20,300 tonnes of TNT. Asked later to explain what he had seen, the physicist Isidor Rabi declared: 'I can't tell you, but don't expect to die a natural death'. The second bomb, a uranium device called 'Little Boy' – 3 m long, 0.61 m wide, and weighing 4.5 tonnes – was dropped on the Japanese city of Hiroshima on 6 August 1945. The third, a plutonium bomb called 'Fat Man' – 3.35 m by 1.52 m and weighing 5 tonnes – was dropped on Nagasaki a few days later. Hundreds of thousands died as a result of blast and heat; many thousands were to die later from radiation poisoning. The atomic age had begun in a most brutal and unforgettable way.

A great blinding light lit up the sky and earth as if God himself had appeared among us. There came the report of the explosion, sudden and sharp as if the skies had cracked, a vision from the Book of Revelation.
*James Chadwick, British physicist, after witnessing the first nuclear explosion at Alamagordo, New Mexico, in July 1945.*

**Left** Victims of the brutal atomic age wait for medical treatment after Nagasaki was hit by a plutonium bomb in August 1945.

**Below** The instant of detonation in an atomic explosion shows the intense heat of the fireball before the mushroom cloud begins to form.

# 3 Taming the dragon

## Harnessing nuclear power

Nuclear fission had put into human hands the power locked up in the atom. The atomic bomb had shown that it could be released explosively, but could it also be dribbled out bit by bit in more manageable amounts and used to generate electricity? The success of the wartime piles built to produce plutonium suggested that it could.

In the 1950s that experience was used to build the first nuclear power plants in Britain, the USA, Canada and the USSR. Credit for the first power-producing reactor is generally given to Calder Hall, in Cumbria on the north-west coast of England, which generated its first current in October 1956. Both the USA and the USSR can claim to have had power-producing reactors earlier, but they were relatively small, producing only a few megawatts of electricity. Calder Hall generated 50 megawatts, enough to supply a sizeable town.

A nuclear reactor is a highly complex machine, built on a massive scale. Its purpose is to establish a nuclear fission reaction and control it – to turn it on, off, up or down. The idea is to create a balance in which, on average, one neutron from each fission goes on to create another fission. If it is more than that, the reactor speeds up; if it is less, it slows down. The necessary balance is achieved with control rods, made of a material that strongly absorbs neutrons. They are used for fine-tuning of the reactor, and also for stopping it in a hurry if anything seems to be going wrong. When the control rods are inserted into the reactor, neutrons are

*The world's first nuclear power station of worthwhile size, Calder Hall in Cumbria, England, started generating electricity in 1956.*

soaked up and cannot contribute to further fissions in the fuel, so the reaction slows down and ultimately stops. When they are withdrawn, fewer neutrons are absorbed, so more are available to cause further fissions and the reactor speeds up.

Reactors also need some way of removing the heat produced by fission, otherwise the whole assembly would heat up and melt. This is done by surrounding the uranium fuel with a coolant, generally water but sometimes gas, which is pumped around so that it takes the heat from the fuel and uses it to raise steam. The steam is then passed through turbines, machines with propeller-like blades, which are forced to turn as the steam expands past them. The turbines then drive the machines which generate electricity.

## Thermal reactors

Finally, most reactors also have a moderator like the pile built by Fermi in Chigaco. By slowing the neutrons down, the moderator reduces the chance that they will escape before they have a chance to cause another fission, because slow neutrons are better at causing fissions than fast ones. Moderators are materials containing light atoms like hydrogen, carbon or, best of all, deuterium. The neutrons bounce from atom to atom, losing energy and slowing down until they are travelling no faster than the atoms in a hot gas. In this condition they are known as 'thermal' neutrons, and reactors using them are sometimes called thermal reactors.

Three materials have been used as moderators in thermal reactors: graphite, ordinary water and heavy water. Calder Hall and successive British-designed reactors use graphite and carbon dioxide gas to cool the reactors. American-designed reactors use ordinary water (which contains hydrogen atoms), both for moderating and cooling. The Canadians have designed a series of reactors which use heavy water as a moderator and coolant. The atoms of deuterium (an isotope of hydrogen) in the heavy water make an ideal moderator because they are light and, unlike hydrogen itself, do not absorb neutrons. The main disadvantage is that heavy water is expensive

> Laymen imagine that the most intractable problem in the development of atomic energy was reactor design. This is far from the truth. The early reactors presented designers with far fewer and far simpler problems than those of the diffusion and chemical plants.
> *Lord Hinton, British engineer, who designed the reprocessing plant at Windscale.*

to manufacture. The Russians have two reactor types. One is similar to the American design, while the other uses a graphite moderator but a water coolant. The reactor at Chernobyl, which blew up so dramatically in April 1986, is of this last type.

The engineering of reactors consists of arranging fuel, coolant, moderator and control rods so that the result is a safe, controllable machine. There is one particular difficulty which does not face engineers in other fields. Once a reactor has started working, the fission reaction produces intense radioactivity, which makes it impossible for anybody to go inside it to make repairs. Unlike a jet engine or a conventional coal- or oil-fired plant, it cannot be taken apart and repaired easily if anything goes wrong. Sometimes repairs can be done by remote control, using TV cameras and robot arms, but they are slow and difficult. As a result, the nuclear designer must assume that once the reactor has started work it will run without trouble for its entire life, which may be as long as thirty years.

Of course, the uranium fuel must be replenished from time to time. Although a little bit of uranium goes a long way – 28 g is equivalent to over 90 tonnes of coal or over 2,700 L of oil – refuelling is necessary. Designers have decided that the best way of arranging the reactor is to have the fuel loaded in slim tubes called fuel elements, usually arranged vertically, with the moderator and coolant surrounding them and filling the spaces between them. When the fuel is ready for replacement, the fuel elements, which are bundled together into fuel assemblies, are pulled upwards out of

*The San Onofre PWR, near San Diego, California, USA, was built on the San Andreas fault line, making it potentially one of the most dangerous nuclear plants in the world.*

the reactor into shielded containers and removed. Fresh fuel assemblies can then be inserted to take their place. A typical reactor of the light-water type might have 200 fuel assemblies, each consisting of many fuel elements. During operation, each fuel assembly will remain within the reactor for up to three years.

The fuel assemblies, the moderator that surrounds them and the structure supporting them, form the core of the reactor. It is here that the heat is generated, and the radioactive products of the fission reaction produced. The reactor core is surrounded by an immensely strong vessel, usually of steel up to 30 cm thick, but sometimes of prestressed concrete. Reactor pressure vessels have to be made with enormous care, since any flaws could develop into leaks or even cause the vessel to explode.

The pressure vessels of light-water reactors are made of steel. In the most successful design, called a pressurized water reactor (PWR), the pressure inside the vessel is 150 times greater than outside. This prevents the water from boiling, even when its temperature is far above normal boiling point. Instead it passes under pressure through a heat exchanger, in which it gives up its heat to a second water circuit, this time unpressurized. The water in this circuit boils, produces steam and drives the turbines. A second light-water design, called the boiling water reactor (BWR) is simpler. It also has a steel pressure vessel, but under lower pressure. The water coolant in the vessel is simply allowed to boil, producing steam to drive the turbines.

The gas-cooled reactors that have been built in Britain use a flow of carbon dioxide gas to remove the heat from the core. The gas then passes through a heat exchanger, where it heats water and raises steam to drive the turbines. The first reactors built used a special magnesium alloy for making the fuel elements and took their name from it – Magnox. Magnox power stations were built in Britain before being replaced with an improved design, the advanced gas-cooled reactor (AGR). The AGR runs at a higher temperature and is physically smaller than the Magnox reactor, while producing a greater power output.

It is the job of the nuclear plant designer to maintain a fission reaction in a situation in which the atoms of U-235 are highly dispersed and heavily outnumbered by atoms of U-238. Only 7 atoms in 1,000 of natural uranium are U-235. The rest absorb neutrons and do not contribute directly to the production of power. One way around the difficulty, as we have seen, is to make the best of the available neutrons by slowing them down and increasing the chances that they will cause fission in another U-235 nucleus and maintain the process. A second approach is to increase the number of U-235 atoms to something closer to 3 per cent. Modern thermal reactors do both, employing moderators and enriched fuel.

## Fast breeder reactors

It is possible to design a reactor without a moderator. If the fuel is enriched and sufficiently concentrated in a small core, the fission reaction can be kept going by fast neutrons. Although they are 400 times less efficient at causing fission than thermal neutrons, the reactor will work if there are enough of them and enough U-235 nuclei. The secret is to pack the core of the reactor into the smallest possible space, with a minimum of neutron-absorbing materials.

Fast reactors are therefore difficult to design, but they do have a compensating advantage. They produce a large number of neutrons, many of which are bound to escape from the core and be wasted, However, these neutrons can be put to use if the core is surrounded by a 'blanket' of U-238, in the form of extra tubes, identical in shape to fuel elements. The atoms of U-238 absorb the neutrons and are converted into Pu-239 which is a fissile isotope. In this way, fast reactors can be used to produce nuclear fuel even as they are consuming it, and they are known as fast breeders. Under ideal circumstances, fast breeders can actually produce more fuel than they consume. This sounds impossible but by turning useless U-238 into useful Pu-239, fast breeders make much better use of the raw uranium. They can in principle produce about 50 to 70 times as much electricity as thermal reactors from a given quantity of fuel.

Why, then, have fast breeders not taken over as the normal type of reactor? First there is the difficulty of designing them. The core of the fast breeder reactor is small compared with other reactors and produces so much heat that cooling it is a major problem. Neither water nor gas is really capable of this and fast breeder designers have all selected liquid sodium, a very good coolant, although a difficult material to handle. The second reason is that the greater economy of the fast breeder would be a decisive factor only if uranium were in short supply, or extremely expensive. For the

*A dummy fuel sub-assembly is lowered into the core tank of Dounreay's prototype fast reactor in Scotland.*

*The world's first commercial-scale fast breeder reactor at Creys-Malville in France, has had serious problems with coolant leakage.*

past thirty years, this has not been the case, so early enthusiasm for fast breeders has tended to diminish. Only one commercial-scale fast breeder has been built, at Creys-Malville near Lyons in France, and it was closed down in May 1987 for extensive repairs after its liquid sodium coolant tank sprang a leak.

**Nuclear propulsion**

Normal nuclear power plants are huge structures weighing thousands of tonnes, but it is possible to design much more compact reactors. In the early 1950s it became clear that submarines powered by nuclear reactors would be able to stay under water for weeks or even months at a time. Conventional submarines use batteries to drive them when they are submerged, so must surface from time to time to run their diesel engines and recharge their batteries. However, nuclear plants can run for months without refuelling – the ideal answer for naval submarines, which must hide for as long as possible beneath the sea.

The design that was chosen and successfully developed for nuclear submarine propulsion was the PWR, using highly enriched fuel to keep the reactor as small as possible. The latest American nuclear submarines, which carry Trident missiles, are huge boats, as long as two soccer pitches. They can spend months on patrol, hiding away in deep water, their engines so silent that it is almost impossible to detect them. They are loaded with nuclear missiles which can be fired from beneath the water. Nuclear propulsion is also valuable for aircraft carriers, which are the main strike weapon of the modern navy. However the complexity, potential danger, and public anxiety about nuclear propulsion have prevented it from being used for commercial vessels, apart from a few experimental ships.

# 4 Nuclear fuel and nuclear ashes

## Uranium sources

Nuclear power stations not only burn a fuel unlike anything used before, but also produce a totally new kind of ash. The fuel begins in the ground as pitchblend, the ore from which uranium is extracted, and finally returns to the ground as nuclear waste. In between, it is involved in a complex cycle, every step of which requires scrupulous care.

*Uranium ore is mined at Mary Kathleen, North Queensland, Australia by opencast methods.*

Uranium ores have been found in many parts of the world, with the richest in the USA, Canada, southern Africa and Australia. Apart from France, which has sufficient uranium for about half its needs, western Europe does not have significant uranium reserves. Finding uranium is much easier than one might expect. Its natural radioactivity, detected in the early days by Geiger counters and today by sensitive instruments carried by helicopters, gives its presence away. The tell-tale sign is the presence in the air of a radioactive gas, radon, which is given off by uranium.

Uranium ore can be mined either by scraping away the surface soil and removing the rock for crushing and processing, or by conventional underground mining techniques. The presence of radon and other radioactive isotopes produced by the uranium ore makes underground mining dangerous: at least 100 American miners who worked in underground mines after the Second World War are known to have died of lung cancer from inhaling radioactive gases and dust. Today much better ventilation has reduced these risks.

The rock containing the ore is first crushed, then dissolved in chemicals to remove the uranium, which emerges in the form of uranium oxide, a yellow substance that is given the name yellowcake. It may be necessary to crush and dissolve 1,000 tonnes of rock to extract 1 tonne of uranium, but the energy stored in the uranium is such that the effort is worthwhile. The uranium is shipped as yellowcake to the fuel fabrication plants.

**Uranium enrichment**

Natural uranium can be used to fuel nuclear plants, as we have seen, but all modern reactors use enriched uranium, in which the proportion of the fissile U-235 isotope has been artificially increased. The enrichment process involves taking advantage of the very small difference in mass between U-235 and the more plentiful U-238. Two techniques have been developed. Both start by converting the yellowcake into a gas called uranium hexafluoride, known as hex. Then the hex is either pumped through very fine filters, or whirled around in centrifuges, using the difference in mass to achieve separation. The first process to be developed, known as gas diffusion, uses thin metal membranes, which allow the smaller molecules, containing U-235, to pass through fractionally quicker than the bigger U-238 molecules. But each membrane increases the concentration of U-235 by only one part in 1,000, so an entire plant consists of a series of thousands of membranes, through which the flow of hex is directed. Ultimately a plant like this can achieve a complete separation, to produce uranium of the quality needed for bombs – more than 90 per cent U-235 – but for nuclear fuel an enrichment of somewhere between 2 and 4 per cent U-235 is all that is needed.

*Yellowcake (uranium oxide) is produced from the ore as the first step in making nuclear fuel.*

*Uranium oxide fuel is compressed into small pellets before being packed into fuel pins.*

The second enrichment technique developed is gas centrifuge. In this process the hex is whirled around in containers at very high speeds, and the heavier molecules, containing U-238, tend to be carried out to the perimeter of the chamber, leaving the lighter U-235 molecules behind. Again, a single centrifuge cannot achieve more than a fractional enrichment, so a whole series, known as a cascade, is needed to reach the target. The entire plant contains many thousands of identical centrifuges, made of lightweight alloys and reinforced with carbon fibre to withstand the forces generated when they are rotated at enormous speeds.

Once enriched, the hex is turned into fuel elements. While some early reactors used uranium metal as their fuel, modern reactors use uranium in the form of its oxide. So the first step is to convert hex into uranium dioxide powder, ensuring the very highest purity and cleanliness of the product. Then the oxide is compressed into small pellets about 1.25cm long and about 1cm in diameter, and heated in a furnace to harden them to the consistency of unglazed pottery. Any faulty pellets are crushed up and sent back to go through the whole process again. Finally the pellets are packed tightly into tubes, held in place with a small spring, and sealed firmly with a cap welded on the end of the tube. A number of tubes are then put together to make a fuel assembly, which is inserted into the reactor.

*Spent fuel is transported inside a canister, surrounded by a very strong shipping flask.*

Once inside the reactor, the uranium atoms start to fission, producing heat and neutrons. In addition, they produce a lot of new elements which are either the result of the splitting of the heavy uranium atom into smaller, lighter atoms or the product of neutron bombardment of U-238. Many of the new elements produced do not exist for long before they themselves break up to form yet further new elements. The result is that within quite a short time the fuel, which consisted initially of uranium dioxide, has been transformed into a very complex mixture of different elements. Some are gases, like krypton and xenon; some vapours, like iodine; and some solids, like caesium, strontium, and the heavy metals produced by neutron capture, including plutonium.

## Nuclear waste

The effect of these new elements is that they interfere with the operation of the reactor, so that long before all the U-235 in the fuel has been consumed, the fuel assemblies must be removed and replaced with fresh ones. What happens to the old, spent nuclear fuel assemblies is a controversial question. In some countries, they are simply treated as waste, while in others they go on to another process in which they are separated chemically into their constituent parts. This is known as reprocessing.

Whatever their ultimate fate, the spent fuel elements are first left for a while to allow their radioactivity to decrease. Among the many fission products present, there are some which decay away very rapidly, while others may take years or even centuries to decay. These differences are measured by a concept known as 'half-life', which is the time taken for any isotope to lose half of its radioactivity. If you started out with 1,000 atoms of a radioactive isotope, the half-life would be the time taken for this number to decline to 500. In another half-life it would decline to 250, in a third to 125, and so on. The half-lives of the isotopes in spent nuclear fuel vary from fractions of a second to tens of thousands of years. But by leaving the spent fuel lying in a pool of cold water close to the reactor for a year or eighteen months, many of the short-lived isotopes will decay, making the spent fuel easier and safer to handle.

It remains, however, very dangerous and must be transported either for final disposal or to a reprocessing plant in a specially designed canister of massive proportions, so that even if there were an accident, it would not result in leakage of radioactivity. In Britain its journey is by road and rail to Sellafield (which used to be called Windscale), a reprocessing plant on the Cumbrian coast. It is right next to the Calder Hall power station.

The reprocessing plant is the dirtiest part of the nuclear fuel cycle. While fuel fabrication and reactor operation produce comparatively little radioactive leakage, reprocessing is

*At Sellafield, spent nuclear fuel is cooled and stored in a pool of water before reprocessing.*

*Shipping flasks containing spent nuclear fuel are delivered to Sellafield from British nuclear power stations on board a special train.*

messy. It involves chopping up the used fuel elements and dissolving them in strong acid, before separating out the constituents into three distinct 'streams'. In one stream is the uranium, including unused U-235, which can be returned to the fuel fabrication factory for manufacture into new fuel elements. In another is the plutonium produced in the reactor when neutrons were captured by nuclei of U-238 to produce Pu-239. In the third is high-level nuclear waste, including the fission products. Because the spent fuel is highly radioactive, all the reprocessing must be done behind thick concrete shielding to protect the operators.

The purpose of reprocessing is to extract any valuable unburned uranium – which can be returned to repeat the process – and the plutonium that has been generated inside the reactor. The first reprocessing plants, built in the USA in the early 1940s, were designed to extract the plutonium which went into the Nagasaki bomb. Plutonium can also be used in nuclear power stations, though so far it has only been used in fast breeder reactors. However, it is potentially a very valuable fuel, too valuable to throw away, according to those who believe in reprocessing. At present plutonium that is not used for making nuclear warheads is stored, in the expectation that one day it will be needed as a nuclear fuel.

A second purpose of reprocessing is to get the highly-active nuclear waste into a form in which it is easier to store. At Sellafield it emerges as a solution of waste products in nitric acid, which is then stored in large, double-walled tanks. But critics say that it would be just as easy, and equally safe, to store the waste inside the original fuel elements without reprocessing. We shall cover these arguments in greater detail in the next chapter.

# 5 What can go wrong?

## The opposition organizes

Nuclear power can be seen as a bargain. It offers virtually unlimited amounts of electricity at economical prices. It does not depend on coal miners risking their lives every day, nor on oil-rich nations consenting to sell their mineral wealth for consumption by others. It does not pollute the air with smoke, or kill the forests and lakes with acid rain. Its record of safety, judged by the bare statistics, is remarkably good; so far it appears to have killed far fewer people than coal mining or drilling for oil.

However, it does exact a price: a vigilance far keener than any other industry has ever demanded. When that vigilance slips, as it did at Chernobyl in April 1986, the consequences are dreadful. A nuclear accident, unlike a major industrial disaster, has consequences that may last for generations. Apart from those killed in the accident itself, there are thousands of others who may die decades later from cancers caused by it. Huge areas of countryside may be laid waste, whole towns and villages made uninhabitable. All this was predicted during the 1950s, when the nuclear age had hardly begun, but the predictions were generally dismissed as the exaggerated fears of a minority who did not understand the technology they were criticizing.

The opposition became more sophisticated during the 1970s and began to tackle nuclear engineers on their own ground. A number of pressure groups began to campaign on the issues of safety, reliability and the disposal of nuclear waste. The first stirrings of organized opposition came from a group in the USA called the Union of Concerned Scientists, who questioned whether nuclear plants were really as safe as their designers claimed. The issue was taken up by Friends of the Earth, an international lobbying group who have become the single most influential anti-nuclear group, appearing at public inquiries and publishing a large number of books and pamphlets on the subject. Nuclear waste and nuclear dumping have been the speciality of another group, Greenpeace, which has also campaigned against nuclear tests.

The effect of these groups has been significant. While in the late 1960s it was hard to find anybody who was critical of nuclear power, by the early eighties opinion had begun to change. The events of April 1986 proved the critics' fears well-founded and contributed powerfully to this change of mood.

*A Greenpeace boat attempts to prevent the dumping of low-level waste at sea by blocking the path of British ship* Gem.

**Left** *Friends of the Earth demonstrate against transportation of spent nuclear fuel.*

## Chernobyl – what went wrong?

Unit number four at Chernobyl, one of more than twenty similar reactors either operating or under construction in the USSR, had always operated well. Moderated by graphite and cooled by water, the reactor was of a type known in the USSR as the RBMK (Graphite Moderated Channel Tube). Many Western experts believe it to be a bad design which can only be operated safely by strictly following the rule book. It is particularly difficult to control at low power, when any small increase in power tends to feed on itself, producing a sharp increase in the activity of the reactor.

The operators at Chernobyl were carrying out an experiement which called for the reactor to be turned down to low power, and then for the steam to the generators to be cut off. The idea was to see how long the turbines would keep spinning under their own momentum and whether they would generate enough power

> The plants are safe: it's the people who aren't.
> *John Kemeny, Chairman of the Commission that investigated the Three Mile Island accident.*

**Below** *Chernobyl's control room before the accident. The display with bright points of light (centre) shows the fuel rod arrangement.*

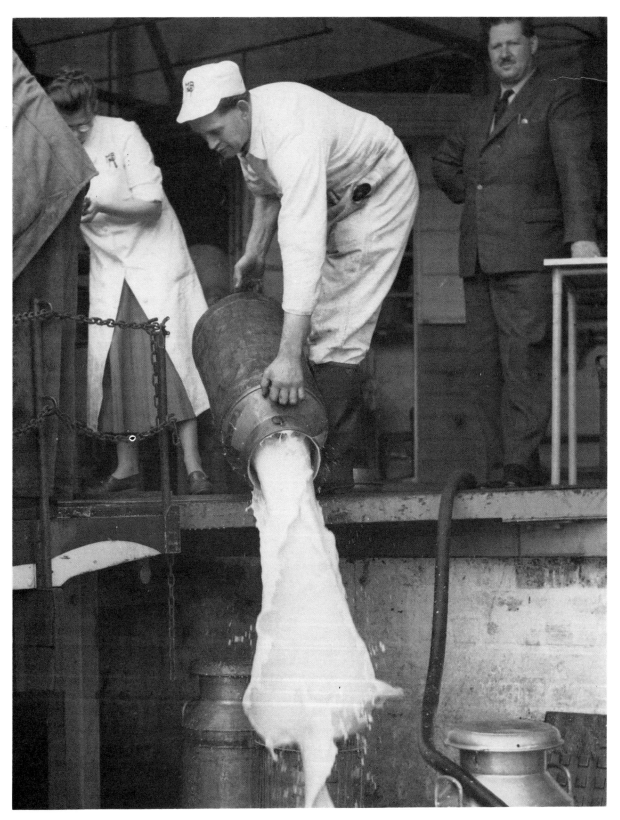

**Above** *Radiation checks on the destroyed reactor at Chernobyl are made by helicopter.*

**Left** *Two million litres of contaminated milk had to be poured away after a fire at Windscale.*

to keep the water pumps working. However, the people in charge of the experiment did almost everything wrong. First, they allowed the reactor's power level to fall dangerously low. Secondly, they overrode the automatic safety devices which would otherwise have closed the reactor down, thus disabling the very systems that could have averted catastrophe. Because reactor power was too low, they pulled the control rods out of the core, and finally they switched off the last safety system, which would have closed down the reactor when steam to the turbogenerators was cut. They did this, in breach of instructions, because they wanted to be able to repeat the experiment if it failed the first time.

This left the reactor running with the control rods out and the safety systems off. As soon as steam to the generators was cut, power rose sharply. The control rods began to slide back into the core, but too slowly to control it. A few seconds before the end, the reactor was 'scrammed' – emergency control rods were inserted – but by then it was too late. In four seconds, power surged to 100 times the full rating, the top was blown off the reactor, the fuel channels overheated and burst, and white-hot particles of fuel were sprayed out, together with burning pieces of graphite. A second explosion followed within a few seconds.

Fortunately, the nuclear reaction ceased almost at once. What was left was a reactor with its lid off, intensely hot and spewing radioactive pollution into the night sky. The extraordinary effort of the local firemen resulted in control of the blaze, thus saving the other three reactors on the same site from destruction. The destroyed reactor remained red hot, however, its graphite core glowing like a barbecue as the fission products escaped. Nobody had ever tackled a reactor disaster as serious as this before; there was no previous experience to draw on. The Russian scientists decided to bury the ruined reactor under a huge mound of materials designed to kill the fire and staunch the leakage of radiation. More than 5,000 tonnes of sand, clay, boron carbide, dolomite and lead were dropped on the reactor from helicopters in hundreds of missions lasting for two weeks. Eventually the amount of radioactivity being released began to fall but by then the damage had been done. About a fifth of the volatile radioactive isotopes of iodine, caesium and tellurium in the core had been released, together with much smaller amounts of strontium, barium, plutonium and cerium.

A total of 31 people died in the accident and its immediate aftermath, and 135,000 had to leave their homes. The settlement of Pripyat,

> Nobody in the world before has ever been confronted with an accident of this kind.
> *Yevgeny Velikhov, chief scientist at the Chernobyl disaster site.*

where many of them lived, is today a ghost town, its blocks of flats eerily quiet and empty. In some windows, washing can still be seen hanging where it was left in the hurried evacuation. In the town, all the trees were heavily contaminated by radiation and have been removed. It is unlikely that the town will ever be occupied again, although the birds, who all disappeared at the time of the accident, had returned by the summer of 1987.

The disaster at Chernobyl was not the first the nuclear industry had experienced but it was by far the worst. In 1957 a serious fire at one of the plutonium-producing piles at Windscale released significant amounts of radioactivity. The National Radiological Protection Board later estimated that the release of radioactivity was likely to have caused 260 extra cases of thyroid cancer and that 33 people were likely to have died of cancer, or sustained genetic damage that would cause disease or death to their descendants. In 1979 a nuclear plant at Three Mile Island, near Harrisburg in Pennsylvania, USA, almost melted down after

> Man has not grown up enough to be trusted with nuclear reactors.
> *Sir George Porter, chemist and Nobel Prize winner.*

a series of errors and malfunctions. Nobody was killed by the Three Mile Island accident, but a lot of people were frightened by the evidence it provided of an industry ill-prepared for anything that might go wrong. A commission of inquiry set up by President Carter of the USA to investigate the accident concluded that the industry was so pervaded by complacency that an accident 'was eventually inevitable'. A total fuel meltdown had only just been avoided at Three Mile Island more by luck than judgement, since the commission concluded that, for most of the duration of the accident, the operators had little idea what to do.

*Houses cluster dangerously close to the Three Mile Island plant in the USA.*

## The health risks

Accidents like these have seriously eroded public confidence in nuclear power, but they are not the only cause for anxiety. There is now growing evidence that children who live close to some nuclear plants have a significantly greater risk of dying from leukaemia, a cancer of the blood, than other children of the same age and background. Initially, scientists found these figures hard to believe, because the nuclear plants involved add only small amounts to the background radiation that already exists in the area. But the evidence is mounting, and it is hard to ignore.

In Britain the evidence is in the form of 'clusters' of cases of childhood leukaemias, centred around nuclear plants. Health statistics show that children under the age of ten living in Seascale, a village near Sellafield, have 10 times as great a chance of suffering from leukaemia as do children in the population as a whole. When these figures first emerged, an expert committee, chaired by Sir Douglas Black, investigated them and concluded that the cluster of cases was real, but that its link with radioactivity could not be proved. It recommended a further study,

*In 1988 a study of the incidence of childhood leukaemia around this plant at Dounreay in Scotland, offered further evidence of the dangers of radiation.*

the results of which were published in June 1988. These demonstrated a further cluster of cases centred around the nuclear establishment at Dounreay in Scotland, where Britain's fast-breeder research is done. The number of cases was very small – just 6 – but only 3 would have been expected from a population this size.

> Nuclear power is the safest form of energy yet known to man.
> *British Energy minister, Peter Walker, a month before the Chernobyl accident.*

Professor Martin Bobrow of Guy's Hospital in London, who led the Dounreay inquiry, commented: 'It seems hard to imagine that the same disease cropping up at Sellafield and Dounreay is just coincidence. To me it seems that the burden of proof has shifted: the leukaemia cases must be assumed to be connected with the nuclear plants until proved otherwise.'

How they are caused remains a mystery, however. The Dounreay plant adds only 1 per cent to the natural radioactivity of the area, so ought to increase leukaemia cases by 1 per cent, not double them. Is there some unknown way in which radiation affects young children, or are there chemicals used in the reprocessing plants at both sites which cause the disease? At present nobody knows.

## Nuclear waste – a legacy for future generations?

One of the unique features of nuclear power, as we have seen, is the waste it leaves behind. The volumes of waste are very small, but a little bit of radioactive material goes a long way and, unlike ordinary rubbish, nuclear waste is not something you can bury in an old quarry and forget about. For nuclear waste remains potentially lethal for hundreds and even thousands of years after it has been produced.

The nuclear industry divides its leftovers into high-, medium- and low-level waste. Low-level waste is only very slightly contaminated and consists of things like laboratory equipment, old gloves, overalls, filters. The danger it poses is not very great, but the volume is quite large: a nuclear power station might produce about forty big drums of such materials every week. In the past they have been disposed of by dumping at sea (now banned by international agreement) or by shallow burial in guarded dumping sites. Growing public anxiety about waste has made a better long-term solution necessary and the idea now is to dump it, together with the slightly more radioactive medium-level waste, in deep burial sites in stable geological formations, where the radioactivity cannot escape to contaminate the ground, water or the air.

The British Nuclear Industry Radioactive Waste Executive (NIREX) had intended to establish shallow burial sites for low-level waste in concrete trenches dug into clay but found that people living in the vicinity of the proposed sites took strong exception to the plans. Preliminary geological surveys were prevented by sit-ins and protests, until NIREX eventually abandoned the search. They have yet to find a suitable site for deep burial.

**Right** *Proposals to test-drill land at Bradwell in Essex, England, for suitability for nuclear waste storage, were opposed by all age groups.*

**Below** *Much low-level waste consists of lightly contaminated protective clothing.*

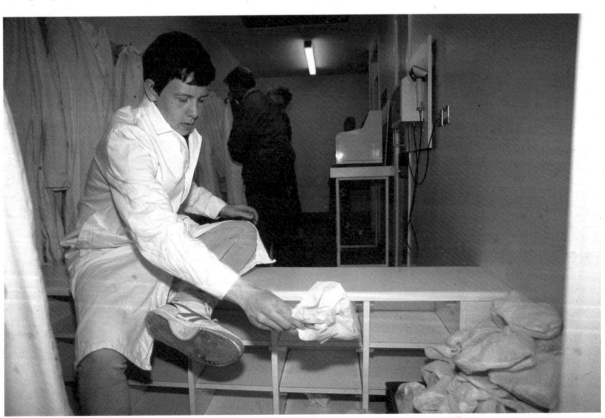

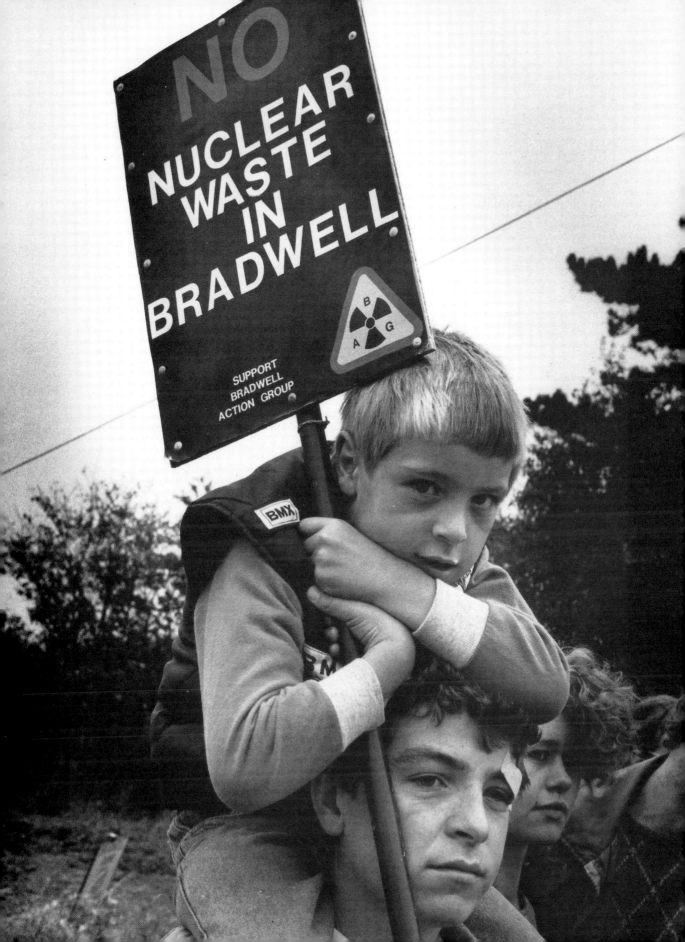

The high-level waste is what is left after reprocessing has separated uranium and plutonium from spent nuclear fuel. In Britain it is stored at Sellafield in large double-walled tanks, cladded with 1.22 m of concrete as protection against the escape of radiation. Because of the heat generated by the fission products as they decay, the dark brown liquid in the tanks has to be cooled constantly to prevent it boiling dry. The industry admits that this form of storage is only temporary, though it has so far lasted for forty years.

In the longer term it is intended to dry the wastes to a powder, then mix them with glass-making materials and fire the mixture in a kiln to produce a glassy solid waste which can be cast into steel cylinders. Such 'vitrified' waste would be easier to handle, less likely to escape, and could be stored safely in deep burial sites. It is already being produced in a plant at Marcoule, France. The waste from a year's operation of a reactor consists of 15 glass blocks, each weighing 350 kg. In this form the wastes continue to generate heat, which is removed by the circulation of air.

Vitrification may provide the long-term answer the industry has been seeking to the problem of high-level wastes. Stored in deep, dry vaults, the waste would gradually lose its lethal properties as the radioactive materials gradually decayed. However, it would take several hundred years to become safe and many thousands of years before it became totally harmless. This is because, as well as fission products, the waste contains traces of plutonium, with a half-life of over 24,000 years. Nuclear waste truly imposes an unusual burden of care on the society that has created it.

*Nuclear waste can be solidified and made easier to handle by turning it into a form of glass.*

# 6 Is a nuclear future inevitable?

To many people opposed to nuclear power, it has seemed at times a hopeless task to halt or even slow its progress. Throughout the 1960s and 70s governments almost everywhere were enthusiastic supporters, seeing nuclear power as both inevitable and good. To build nuclear power stations was evidence of a modern, go-ahead approach; to oppose them was to adopt the tactics of the Luddites, who smashed the new textile machinery of the Industrial Revolution in Britain because they believed it threatened their jobs.

## The nuclear tide turns

Today the situation is different, however. Some countries, responding to the pressure of their citizens, have decided to build no new nuclear power stations, though continuing to use the ones already in service – Sweden and Italy are examples. Other countries, like Austria and the Philippines, have gone even further, declaring that they will not bring into service plants already complete. In the USA, where electricity generation is in the hands of private companies

> The first lesson we've learned is 'Don't build nuclear plants in America'. You subject yourself to financial risk and public abuse.
> *Electrical utility manager, quoted in* Newsweek *in 1984.*

*Shoreham plant on Long Island was abandoned before it generated a single watt of electricity.*

independent of government, no new orders for nuclear plants have been placed since 1979. The companies have decided that public opposition is too strong and the likely legal cost of defeating it too high. One company, in Long Island, New York, has even had to abandon a complete plant which had cost £3 billion to build and was ready to start, because local opponents were worried about how they would escape from the area down the single main road off the island in the event of an accident. After years of court proceedings, the Long Island company finally admitted defeat and abandoned its expensive plant. All over the USA, citizens are making it clear that they would rather pay more for their electricity than live close to a nuclear plant.

Yet at the same time other countries are pushing on with nuclear power. France now generates almost two-thirds of its electricity from nuclear plants and is continuing to invest heavily in them. The British government is also strongly committed and has said that when the electricity industry is privatized, the private companies will not be allowed to turn their backs on nuclear power, as they have done in the USA. The bigger of the two companies into which the present Central Electricity Generating Board (CEGB) is to be split will be given responsibility for the existing nuclear stations and for building new ones, including a PWR at Sizewell in Suffolk. The USSR also remains committed to nuclear power, despite the experience of Chernobyl.

Nevertheless, it is quite clear that the prospects for nuclear power have changed dramatically for the worse in the past decade, mainly because of the accidents at Three Mile Island in 1979 and at Chernobyl in 1986. But these accidents would not have had such an effect on public opinion if there had not already been doubts about nuclear power. They were of three kinds: environmental, political, and economic.

The environmental arguments centre around the safety and cleanliness of nuclear plants. It is one thing, critics argue, for nuclear stations

*Demonstrators in the Philippines were effective in bringing about the cancellation of an American nuclear plant.*

to be built in countries with a highly qualified workforce, an established engineering profession and a record of professional integrity in the public services. The people who design the stations can be trusted, while those who operate them will do their jobs well. But could the same be assumed if hundreds of nuclear plants were to be built all over the world? Many people, even inside the industry, doubt this. The accident at Chernobyl has tended to confirm their doubts. Both the design and operation of the reactor that blew up so dramatically can now be seen to have been deficient.

Beyond this there is a wider argument about the uses to which nuclear reactors may be put. As we have seen, the nuclear reactor and the nuclear bomb are Siamese twins. The same reactors that generate electricity can be used to produce plutonium from which weapons can be made. From the very beginning of civil nuclear power – the 'Atoms for Peace' initiative launched in the USA by President Eisenhower at the United Nations in 1953 – it has been recognised that the spread of nuclear power might be accompanied by the simultaneous spread of nuclear weapons. That was the last thing Eisenhower wanted, so he proposed an international agency to control the technology: the International Atomic Energy Agency, guided by the United Nations (UN). In 1968 the USA and the USSR presented to the UN the text of a treaty they had drafted to control the spread of nuclear weapons: the Nuclear Non-proliferation Treaty (NPT).

The idea behind the NPT was that the developed countries would help the less developed ones to benefit from nuclear power, but only if they agreed not to use it as a source for their own weapons. Many countries saw this as an attempt by those who had nuclear bombs (at that time the USA, USSR, and Britain) to deny the ultimate weapon to those without it and therefore declined to sign. Among these were France, China and India, which were already developing their own weapons, and a host of other countries which have not yet developed nuclear weapons but do not want to lose the option: Argentina, Brazil, Pakistan, Saudi Arabia, South Africa, North Korea, Vietnam and Zambia, to name just a few. In

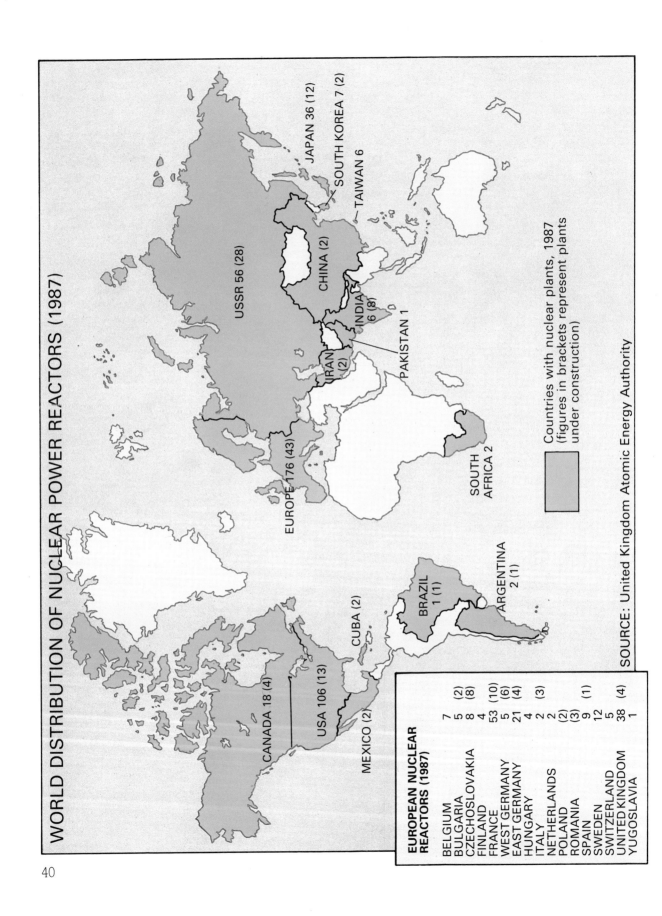

total about a quarter of the world's nations have not agreed to the terms of the NPT.

Being aware that the NPT might fail, the main exporting nations established a less formal mechanism in the 1970s: the London Nuclear Suppliers Group. It agreed a list of sensitive technologies that should not be exported without stringent safeguards, but it was difficult to pretend that nuclear proliferation could be prevented forever by such means. After all, there was nothing to stop any country from learning all it needed through importing safeguarded technology, then setting up its own domestic industry and building a bomb on its own. This was the course followed by India (whose co-operation with Canada was brought to an end with the explosion of the first Indian nuclear device in 1974) and more recently by Pakistan. Israel is another country which has gone its own way and is now believed to have as many as 100 nuclear weapons, though it has never actually tested one.

American policy in the 1970s discouraged exportation of reprocessing plants (the key to proliferation because they provide the means of extracting plutonium from spent fuel). After he was elected president of the USA in 1976, Jimmy Carter even stopped reprocessing of spent American fuel, except for nuclear weapons production. His view was that the USA should not make use of what it denied others and that reprocessing was too dangerous a technology for a world of nation-states frequently in disagreement with each other.

The third and most decisive blow against nuclear power has been economic. In the early years the claim was made that electricity generated by nuclear power would be far cheaper than from any other source, 'too cheap to meter' in the words of one enthusiastic advertising man. The main cost in nuclear generation is building the plant, not supplying the fuel. Plants burning coal or oil are much cheaper to build, but the cost of fuelling them is higher. When in 1973 and again in 1979 the cost of oil was sharply increased by the actions of the Organization of Petroleum Exporting Countries (OPEC), the balance appeared to have shifted even more heavily towards nuclear power. Oil became too expensive to burn in power stations; coal was dependent on the goodwill of miners. Nuclear power was viewed as the energy source of the future.

Surprisingly, however, it did not work out that way. One of the main effects of the 'oil shocks', as they were called, was to create worldwide inflation. Among the sectors which suffered worst from inflation were large projects like power stations. The cost of building them soared. Growing concern about the environment, which forced designers to add extra safety systems to nuclear plants, increased the costs still further. The unexpected result was that, although coal had increased in price, nuclear power costs had increased just as much. So the bonanza of new orders that was expected did not come about, except in France, which made a conscious decision to build a nuclear-based economy and stuck to it.

The inflation and economic depression that followed the oil shocks had another effect which damaged nuclear power. The demand for electricity, which had been rising strongly for the previous thirty years, went temporarily into decline. The huge increase in costs made people more careful about how much they used, and the decline in manufacturing industries also cut demand. As a result, the electricity companies did not need to build new plants. On the contrary, they could afford to close down their older ones and concentrate generation on the most economic. Orders for new plants dried up almost completely. Today, with the demand for electricity recovering, new orders are being placed, but will they be for nuclear power?

In Britain, the CEGB are to close a Magnox power station at Berkeley, Gloucestershire, because it would cost too much to bring its safety systems up to modern standards. One new nuclear station has been ordered to be built at Sizewell: a PWR, the first light-water reactor to be built in Britain and the subject of a hard-fought public inquiry. The final report in January 1987 recommended that, despite some risks to health and to the environment around the plant, the national interest would be served by building the PWR at Sizewell as the first in a series of British PWRs.

The situation in the USA is less than clear-cut. Given the very lengthy licensing procedures that are now necessary for new plants, it can take ten to twelve years from the moment of decision until the first watts are generated, compared with less than five years in France. This makes it very difficult for the electricity companies to raise the money, when the financial return is so far off. This, coupled with public opposition, has led to American utilities being reluctant to order nuclear power. So far, there are no signs of a real recovery in the industry, though it has sufficient work already in hand to keep it going.

In Europe the spectrum runs from France at one extreme to Austria at the other. The French investment has been enormous: 33 PWR plants were built between 1975 and 1985, with a further 22 expected to be completed by 1990. The nationalized electricity industry, Electricité de France, has run up colossal debts to

> The problem of the use of nuclear energy in Austria can be considered as decided and closed.
> *Austrian Chancellor, Fred Sinowatz, announcing to parliament that Austria's only nuclear plant would be dismantled.*

pay for all these plants – more than $200 billion – making it one of the biggest debtors in the world. At the other extreme is Austria, which built a single boiling-water reactor at Zwentendorf, close to the Czechoslovakian border, before holding a referendum in 1979 in which the people voted against nuclear power. The reactor was mothballed and finally, following

*The river Loire, in France, long famous for its historic chateaux, now has several nuclear power plants along its banks as well.*

the Chernobyl disaster, the decision was taken to dismantle it.

Given the wide differences between countries, it is impossible to predict the future for nuclear power. Optimists in the industry say that, as demand for electricity recovers, people will overcome their aversion: given a choice between nuclear power and no power at all, they will accept the atom for all its risks. Critics, on the other hand, say that the future of nuclear power has been fatally damaged by the accidents at Three Mile Island and Chernobyl and can only survive in countries where the government imposes it on a reluctant population. Either of these predictions may be right; at the moment it is simply too soon to say.

A judgement would be easier to make if there were any real and favourable alternatives to nuclear power in sight. After the first oil shock, considerable effort was put into developing alternative sources of energy: solar, tidal, wind, geothermal. All have possibilities but at the moment none seems likely to become a really major source of power in the near future.

The same is true of another long-term hope, nuclear fusion. Unlike fission, in which a heavy atom splits to form lighter ones, fusion is the process in which light atoms combine to form heavier ones. It is the reaction that provides the energy of the sun and of the hydrogen bomb. Taming it has so far proved much more difficult than controlling fission, however. In spite of more than thirty years of effort, a fusion power station still seems a long way off. If one could be built, it would have distinct advantages, including much lower levels of radioactivity. However, while it is worthwhile continuing research, nobody should count on fusion coming to the immediate rescue.

Nuclear power is always going to raise strong feelings. Its future depends on three things: how well and safely it can perform, how strongly the demand for electricity grows over the next few decades and the extent of public opposition, none of which can be predicted with any certainty. All that seems certain at the moment is that the issue of nuclear power will not go away. The arguments over it are likely to flourish for the foreseeable future.

> It is another tolling of the bell, another grim warning that the nuclear era necessitates new political thinking and a new policy.
>
> *Mikhail Gorbachev describing the Chernobyl accident on Soviet television, 14 May 1986.*

*Has the Chernobyl disaster fatally damaged the future of nuclear power? At present it is too soon to say.*

# Glossary

**Allies**  The countries that fought against the Axis and Japan in the Second World War, in particular Britain and the Commonwealth countries, the USA, USSR, France and Poland.

**Atoms**  The constituents of which matter is made.

**Axis**  The alliance of Nazi Germany, Fascist Italy and Japan from 1936 until their defeat in the Second World War.

**Control rods**  Rods which are made of a material that can absorb neutrons. They are used to control fission.

**Core**  The heart of a nuclear reactor, containing the fuel assemblies and moderator. It is here that the fission chain reaction takes place.

**Critical mass**  The minimum amount of fissile material that is needed to create a nuclear weapon.

**Deuterium**  A naturally occurring isotope of hydrogen, twice as heavy as ordinary hydrogen.

**Electron**  One of the particles found in an atom. It has a negative electrical charge.

**Element**  A distinct kind of atom. There are almost 100 different kinds and these elements make up all known matter.

**Enrichment**  The process of increasing the proportion of uranium 235 in natural uranium.

**Fast breeder reactor**  A reactor which uses fast neutrons to 'breed' more fuel than it consumes.

**Fissile**  Capable of undergoing nuclear fission.

**Fission**  The splitting of the nucleus of an atom into two or more parts.

**Fuel element**  The nuclear fuel sealed into a tube which generates heat in the core of a reactor.

**Fusion**  The process in which two small atoms fuse together.

**Isotopes**  Different forms of the same element. The nuclei have the same number of protons in them, but different numbers of neutrons, so they have different weights.

**Magnox**  A magnesium alloy used to make the tubes of the fuel elements in Magnox reactors.

**Moderator**  A substance used to slow down neutrons during fission.

**Neutron**  One of the particles in the nucleus of an atom.

**Radiation**  The energy given off by atoms, particularly during nuclear decay.

**Radioactivity**  Radiation from atomic nuclei.

**Reprocessing**  The treatment of spent fuel from a nuclear reactor used to separate unused uranium and plutonium from radioactive fission products.

**Thermal reactor**  A reactor in which the neutrons from fission are slowed down by a moderator.

**Transmuting**  Changing an element.

**Yellowcake**  Uranium ore which has been refined.

# Books to read

Chudleigh, R. and Cannell, W. *Radioactive Waste: The Gravedigger's Dilemma* (Friends of the Earth, 1985)
Clark, Ronald W. *The Greatest Power on Earth* (Sidgwick and Jackson, 1980)
Driscoll, Vivienne *Focus on Nuclear Fuel* (Wayland, 1985)
Hamman, Henry and Parrott, Stuart *Mayday at Chernobyl* (New English Library, 1987)
Hawkes, Nigel *Nuclear Power* (Observer Schools Service, 1987)
Lambert, Mark *Focus on Radioactivity* (Wayland, 1989)
Patterson, Walter C. *Nuclear Power* (Pelican, 1983)
Patterson, Walter C. *The Fissile Society* (Earth Resources Research, 1977)
Rippon, Simon *Nuclear Energy* (Heinemann, 1984)

# Picture acknowledgements

Format 42; Greenpeace cover, 27; Network 25, 26, 28, 34, 35, 43; Photri 12, 13, 14 (top and bottom), 22, 23, 24, 36; Popperfoto 8, 30; Rex Features frontispiece, 6, 7, 17, 31, 32, 33; Frank Spooner 20, 29, 37, 38; Topham 9; United Kingdom Atomic Energy Authority 15, 19, 21. The map on page 40 and illustration on page 10 were provided by Peter Bull.

# Further information

If you wish to find out more about some of the subjects covered in this book, you might find the following addresses useful:

Atomic Energy Control Board
PO Box 1046
Ottawa
Ontario
Canada

Australian Atomic Energy Commission
Australian Nuclear Science and Technology Organization
Lucas Heights Research Laboratories
New Illawarra Road
Menai
New South Wales 2234
Australia

British Nuclear Fuels plc
Information Services
Risley
Warrington
Lancashire
England
Tel. 0925 832000

Central Electricity Generating Board
Department of Information and Public Affairs
Sudbury House
15 Newgate Street
London
EC1A 7AU
England
Tel. 01 634 5111

Friends of the Earth Trust Ltd
26–28 Underwood Street
London
N1 7JQ
England
Tel. 01 490 1555

Greenpeace
124 Cannon Workshops
West India Dock
London
E143 9SA
England
Tel. 01 515 0275

Greenpeace Australia
1/787 George Street
Sydney
New South Wales 2000
Australia
Tel. 2 2110089

Greenpeace Canada
427 Bloor Street West
Toronto
Ontario
M5S 1X Southern
Canada
Tel. 922 3011

Greenpeace New Zealand
5th Floor
Nagel House
Court House Lane
Auckland
New Zealand
Tel. 9 31030

Institute of Nuclear Science
Gracefield Road
Lower Hutt
New Zealand

United Kingdom Atomic Energy Authority
11 Charles II Street
London
SW1Y 4QP
England
Tel. 01 930 5454

# Index

The numbers in **bold** refer to the pictures.

Accidents 6, 27–32, 43
Africa, southern 21
Alamagordo, New Mexico 13, **14**
Atom
   discovery of 8
   splitting of 10–12
Australia 21, **21**
Austria 37, 42–3

Becquerel, Henri 8–10
Bombs, atomic 11–14
Britain
   commitment to nuclear power 39
   disposal of waste in 25
   thermal reactors in 16–18
   leukaemia 33
   power plants built in **15,** 15, **33,** 33

Cadmium 13
Calder Hall **15,** 15, 16, 25
Canada **7,** 15, 16, 21, 41
Carter, Jimmy 32, 41
Central Electricity Generating
   Board (CEGB) 39, 41
Chernobyl **6,** 6-7, 16, 27, **29, 31,** 29–32, 39, 43
China 39
Costs 41
Creys-Malville fast breeder 20, **20**
Critical mass 11
Curie, Marie **8,** 10

Dounreay **33,** 33

Einstein, Albert 12
Eisenhower, Dwight 39
Electricity 7, 15–16, 18, 27, 37–41
Electrons 10

Fermi, Enrico **12,** 12–13, 16
Fission, nuclear 11, 15–17
France
   commitment to nuclear power 7, 39, 42
   power plants in 20, **20, 42**

nuclear waste in 36
refusal to sign treaty 39
uranium reserves in 21
war with Germany 11

Friends of the Earth 27, **28**
Fuel, nuclear 21
Fuel assemblies, in reactors 17–18, **19,** 23
Fuel pellets 23
Fusion, nuclear 43

Gas centrifuge, and uranium
   enrichment 23
Germany 11
Graphite 12, 16, 29
Greenpeace 27, **27**

Hahn, Otto 11
Half-life 25
Health risks 33
Hiroshima 13
Hitler, Adolf 11, 12

India 39, 41
International Atomic Energy Agency
   (UN) 39
Isotopes 11
   half-life of 25
Israel 41
Italy 7, 37

Leukaemia 33
London Nuclear Suppliers Group 41

Magnox power stations 18, 41
Manhattan Project 12
Meitner, Lise 11
Moderators, in nuclear reactors 16–18

Nagasaki 13, **14,** 26
National Radiological Protection
   Board 32
Neutrons 10, 11, 15–16, 18, 24
Nuclear Industry Radioactive Waste
   Executive (NIREX) 34

Nuclear Non-proliferation Treaty 39–41
Nuclear propulsion 20

Pakistan 39, 41
Philippines 37, **38**
Plutonium 24
   in nuclear waste 36
   in nuclear weapons 26, 39
   Pu-239 12–13, 18
Pressure vessels 18
Protests **7,** 27, **27, 28, 35, 38, 43**
Protons 10

Radioactivity
   from Chernobyl 6, 31
   discovery of 8–10
   and fusion 43
   and leukaemia 33
   reduction in 25, 36
   and reprocessing 26
   in thermal reactors 17
   in uranium ore 21
   from Windscale **30,** 32
Reactors, nuclear
   fast breeder 18–20
   first **13,** 15–16
   gas-cooled 18
   thermal 16–18
Reprocessing, of nuclear waste 24–6, **25,** 41
Roentgen, Wilhelm 8

Safety 27–36
San Onofre **17**
Sellafield 25, 26, **26,** 33
   *see also* Windscale
Shoreham **37**

Sizewell 39, 41
Submarines 20
Sweden 37

Three Mile Island **32,** 32, 39, 43

Union of Concerned Scientists 27
Uranium
   enriched 22–3
   fuel 17–18
   hexafluoride 22
   ores **21,** 21–2
   radioactivity 8–10
   reprocessing 26
   splitting of atom 11
   U-235 11–13, 18, 22–4
   U-238 11–13, 18, 22–4
USA
   accidents in 32
   decline in nuclear power 37–9
   future of industry in 42
   nuclear power plants in 15, 16, **17,** 32, **32, 37**
   reprocessing plants in 26
   uranium ores in 21
USSR 6, 15, 17, 29, 39

Vitrification, of nuclear waste 36, **36**

Waste, nuclear 21, **24,** 24–6, **34,** 34–6
Weapons, nuclear 7, 11, 39
Windscale 25, 30, 32
   *see also* Sellafield

X-rays 8–10, **9**

Yellowcake 22, **22**